I WANT TO BE A...

FIREFIGHTER

DOUG BRADLEY

New York

Published in 2023 by The Rosen Publishing Group, Inc.
29 East 21st Street, New York, NY 10010

First Edition

Editor: Caitie McAneney
Book Design: Rachel Rising

Photo Credits: Cover, pp. 1, 17 VAKS-Stock Agency/Shutterstock.com; pp. 4, 6, 8, 10, 12, 14, 16, 18, 20 april70/Shutterstock.com; p. 5 wavebreakmedia/Shutterstock.com; p. 7 Firefighter Montreal/Shutterstock.com; p. 9 Arisha Ray Singh/Shutterstock.com; p. 11 Stacey Green/Shutterstock.com; p. 13 ARM Photo Video/Shutterstock.com; p. 15 Gorgev/Shutterstock.com; p. 19 Henrik Dolle/Shutterstock.com; p. 21 Nomad_Soul/Shutterstock.com.

Some of the images in this book illustrate individuals who are models. The depictions do not imply actual situations or events.

Library of Congress Cataloging-in-Publication Data
Names: Bradley, Doug.
Title: Firefighter / Doug Bradley.
Description: New York : PowerKids Press, 2023. | Series: I want to be a... | Includes glossary and index.
Identifiers: ISBN 9781725339811 (pbk.) | ISBN 9781725339835 (library bound) | ISBN 9781725339828 (6pack) |
ISBN 9781725339842 (ebook)
Subjects: LCSH: Fire fighters–Juvenile literature. | Fire fighters–Vocational guidance–Juvenile literature.
Classification: LCC HD8039.F5 B656 2023 | DDC 363.37–dc23

Manufactured in the United States of America

CPSIA Compliance Information: Batch #CSPK23. For Further Information contact Rosen Publishing, New York, New York at 1-800-237-9932.

CONTENTS

What Do Firefighters Do?

Firefighters are important community helpers. While most people run from fires, firefighters run to them. They save lives by putting the fires out. Firefighters are **first responders**. They also help people in car crashes and other **emergencies**.

At the Firehouse

Firefighters usually start their workday at a firehouse, or fire station. This is where all the firetrucks and **gear** are kept. Some firefighters stay at the fire station for a whole day. They eat, sleep, and wait for calls for help.

54
277
CENDIE
339

It's an Emergency!

People call 911. Then firefighters jump into action! They go to house fires, building fires, and forest fires. They also go to emergencies such as car crashes and reports of people who are hurt. They know what to do in all emergencies.

Suit Up!

Firefighters have only a minute to get their special gear on. They wear clothes that can't catch on fire. They wear helmets with face coverings. They wear gloves and boots. They also wear a **tank** with fresh air on their back.

CHANGE

On the Firetruck

Firetrucks also have special gear. The gear helps firefighters put out fires. Firetrucks have powerful hoses that pump water onto fire. They have ladders to help firefighters reach high places. They have tools to help people out of tight places.

FIRE & RESCUE NSW
FIRE
FIRE & RESCUE
NEW SOUTH WALES
FIRE & RESCUE
NEW SOUTH WALES
SEV
BC 12 HP

First on the Scene

Firefighters are often the first people who show up at an emergency. They know how to save a person's life. They know what to do if someone is hurt. They know how to get people to safety.

Saving Lives

Firefighters can save lives every time they get a call. People need to get out of burning buildings—fast! Fire can burn them. Too much smoke can hurt their bodies. Some people need help to breathe. Some people are scared. Firefighters help them.

How to Be a Firefighter

Firefighters go to school and sometimes **college**. Then they go to a fire academy, or school, to learn more. They need to learn how to help a person who is hurt. They need to take special tests. They also need to be fit.

Future Firefighter

Do you want to be a firefighter? You can do it! First, you can learn about fire safety. Firefighters often visit schools to teach people how to stay safe. Listen to firefighters. Maybe you are a future firefighter!

GLOSSARY

college: A school people can go to after high school.

emergency: An unexpected and often unsafe situation that calls for action right away.

first responder: A person responsible for going to an accident or emergency first.

gear: Clothing or tools that are used to do something.

tank: Something that can hold something else.

FOR MORE INFORMATION

BOOKS

Anderson, AnnMarie. *Meet a Firefighter!* New York, NY: Children's Press, 2021.

Kawa, Katie. *A Day with a Firefighter.* New York, NY: Cavendish Square Publishing, 2021.

WEBSITES

So You Want to Be a Firefighter?

inspiremykids.com/careers-that-count-so-you-want-to-be-a-firefighter/

Learn more about what firefighters do and how you can be one too.

What to Do in a Fire

kidshealth.org/en/kids/fire-safety.html

Knowing what to do is important in case of a fire. Look at tips for how to get away safely.

Publisher's note to educators and parents: Our editors have carefully reviewed these websites to ensure that they are suitable for students. Many websites change frequently, however, and we cannot guarantee that a site's future contents will continue to meet our high standards of quality and educational value. Be advised that students should be closely supervised whenever they access the internet.

INDEX